Space Suit Evolution
From Custom Tailored To Off-The-Rack

Apollo Space Suits Were Custom Tailored

The Apollo space suit was basically a one-piece suit. Each suit was made to fit (custom tailored) each astronaut. Each Apollo mission required fifteen (15) suits to support the mission. The main, or prime, three-man crew each had three suits: 1 for flight; 1 for training; and 1 as a flight back-up in case something happened to their flight suit, thus a total of 9 suits for the prime crew. The back-up three-man crew each had two suits: 1 for flight and 1 for training. The astronaut corps at that time included between 25 and 27 astronauts.

Shuttle Space Suits Are "Off-The-Rack"

The Shuttle astronaut corps includes about 120 men and women. The Shuttle space suit, to accommodate the large number of astronauts with widely varying body sizes, was designed to be made up of many interchangeable parts. These parts (upper and lower torso's, arms, etc.) are fabricated at ILC in different sizes, inspected/tested, then shipped to Johnson Space Center (JSC) where they are inventoried for the astronaut corps.

ILC Dover has a staff of about 15 people who work on-site at JSC. This staff is primarily responsible for the control, use, and maintenance of the suit components produced in Frederica after they arrive at JSC. The staff also develops and executes the crew training schedules. This involves everything from taking measurements and conducting fitchecks to de-stow and post-flight inspection/test of the space suit.

Measuring Up For Shuttle

At JSC, the body measurements of each Shuttle astronaut are taken and recorded, then the measurements are plotted against the size ranges available for each space suit component. The suit components are then assembled and the astronaut is "fitchecked." Once the astronaut is satisfied with the fit, then a training suit is assembled using duplicate sized components. The astronaut will use this assembly for many mining events. A flight suit is then assembled using the same size components and the astronaut will "checkout" this suit during chamber testing and other test events.

Training suits are usually assembled nine months prior to flight and flight suits are usually assembled four months prior to flight.

Pack Your Bags!

Just before a Shuttle mission, the suits designated for flight are tested, cleaned, and packed at JSC. Then they are flown to Kennedy Space Center (KSC) and stowed on the orbiter. After each flight the suits are returned to JSC for post-flight processing and reuse.

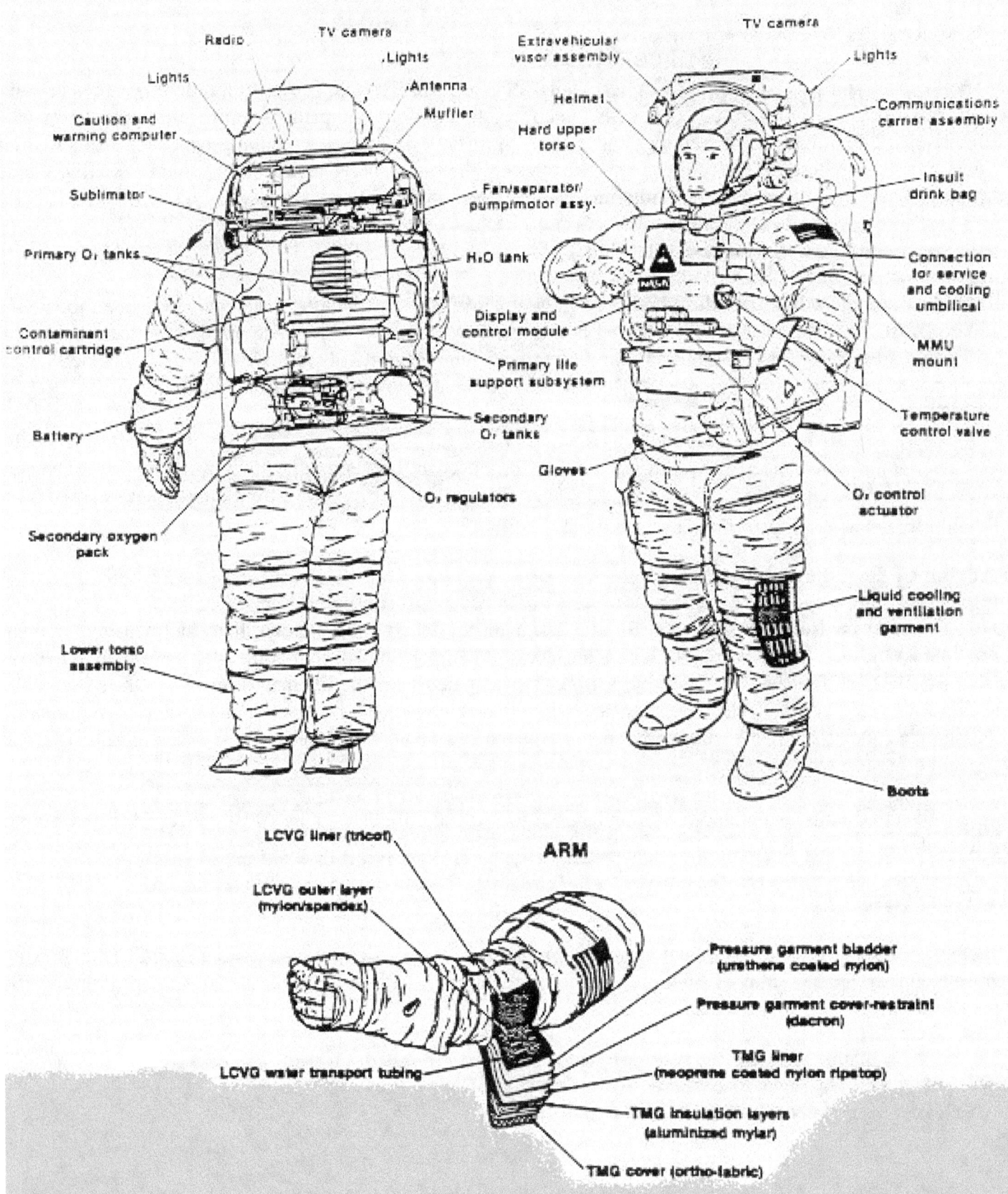

Questions & Answers About Space Suits

1. **I would like to know if you are making old stuff into new stuff?**

   No. Materials used in making the space suits are purchased from other manufacturers and we begin building from scratch. The Apollo suits that were on the moon were designed for a special purpose and were tailored for each astronaut. The new suits used on the Shuttle are not tailored and are designed for special features that were not required for the Apollo suit.

2. **When did you start making astronaut suits?**

   ILC started designing suits in 1961; started making test and prototype suits in 1964; and started delivering suits for use by the Apollo astronauts in 1966.

3. **How thick is the space suit?**

   Approximately 3/16" thick, 11 layers of materials.

4. **Do you have astronaut suits there? How many do you have? How many sizes do you make? Do you make space suits by the dozen, one at a time, or what?**

   ILC no longer builds custom-made suits for each astronaut. We now build separable components (arms, legs, boots, etc.) that attach together which provides many sizes. After a suit has completed its mission, it is disassembled. Some of its parts are then mated with other parts to build another suit of a different size. Once these component parts have been made, they are shipped to NASA's Johnson Space Center near Houston, Texas. We do have a certification suit here, which is manned (worn) at ILC during a space walk. The number of components made depends upon the number of flights scheduled at a given time and the number of people that would wear a particular size.

5. **How do you make astronaut suits and what material do you use?**

   The suits are basically made by sewing and cementing various materials together, and then attaching metal parts that let you join the different components together. Suit materials include: ortho-fabric, aluminized mylar, neoprene-coated nylon, dacron, urethane-coated nylon, tricot, nylon/spandex, stainless steel, and high strength composite materials.

6. **How much does it cost to make a space suit?**

   It is difficult to express the cost of the suit; not all parts are made by ILC, It has been reported in the news media that the spare suit costs two million dollars. This does not mean that each time the suit is used it costs two million dollars. If you would never use the parts for another astronaut or another mission, then you would have a one-time use and the cost could probably be as high as what the news media says it is.

7. **How long does it take to make a space suit?**

If all the parts and materials were available for use by one person. who possessed all the different skills required, it would take that person almost 2-1/2 years to produce a space suit (roughly about 5,000 man-hours of work). With the complement of personnel and various skills at ILC Dover, a suit can be produced in three months.

8. **What color do they have to be?**

The reason that the suits are white is because white reflects heat in space the same as it does here on earth. Temperatures in direct sunlight in space can be over 275 degrees Fahrenheit.

9. **When the astronauts go into space once and then when they go again, do they wear the same space suits?**

The astronauts will wear the same size component parts as the first mission unless body dimensions have changed (for example, gained weight).

10. **How many space suits have you made?**

ILC has manufactured enough separable components to assemble 51 individual Shuttle space suits. Some of these suits are used for the astronauts' training here on earth. These training suits are not intended for flight.

11. **Have you made any changes in the suits?**

Since ILC started developing prototype space suits in 1964, there have been many, many changes based on the usage of the suit. The Apollo suit, for example, was designed for only one mission and it had to be lightweight to allow the astronauts to do work on the moon. The Shuttle suit, however, is designed only to work in zero gravity where the astronaut does not feel the weight of the suit, and it is designed to last for up to 15 years on many missions.

For these reasons, the Shuttle suit is much heavier than the Apollo suit was. The Apollo suit, including the life support backpack, weighed about 180 pounds. The Shuttle suit, including the life support system, weighs about 310 pounds. The suit itself weighs about 110 pounds. If an astronaut weighing 175 pounds wears the complete suit, the total weight is then about 485 pounds (310 + 175 =495).

Changes to the design of the space suit are constantly happening as we learn more about living and working in space.

12. **Is the suit too heavy for training on earth?**

    When training on earth, the astronauts use special lightweight parts and they do a lot of training in water (a large pool called a Weightless Environment Test Facility). When the suit is pressurized with breathing air and the astronaut goes into the water, it is something like being in a balloon and floating on the water. Weights are added around the waist so the astronaut will sink and stay under water. This gives the same feeling of weightlessness the astronaut would feel in space.

13. **How did you find out what cloth to use for the suits?**

    a) By engineers establishing design limits (how much heat and cold the hostile environment of space presents).

    b) Again, by engineers calculating how well various materials will withstand the above established design limits.

    c) By engineers calculating how long the materials will last when the materials are sewn/cemented together, folded and creased, and moved to different positions by bending or rotating the shoulder, elbow, wrist, waist, hip, knee, and ankle.

    d) By many hours of testing material samples.

14. **Do you have to make changes in the suits for women astronauts?**

    There is no difference in a male or female suit, though the female usually requires a smaller size.

15. **How many years has your business been open?**

    Since 1947.

16. **How many people do you have working for you?**

    We have approximately 400 employees in Frederica, Delaware and 15 employees that work at the Johnson Space Center near Houston, Texas. About 60 of the Frederica employees am involved in the space suit program.

17. **Who was the first space suit for?**

    The first suit worn in spare was made for astronaut Alan Shepard when he flew on May 5, 1961 during Project Mercury. This suit was not made by RZ. The first suit worn on the moon was made by ILC for Apollo astronaut Neil Armstrong's July 20, 1969 moonwalk. The first EVA using suits for the Space Shuttle was performed by astronauts Story Musgrave and Donald Peterson on April 7. 1983.

18. How much air does the space suit hold?

The amount of air in the suit will vary, depending upon the size of the suit. The extra large Shuttle suit, without anybody in it, holds 5.42 cubic feet of air; the extra small size holds 4.35 cubic feet. With an astronaut in the suit, the amount of free space remaining in the suit is 2 cubic feet.

19. How long does 2 space walk- (EVA) last?

During an Extravehicular Activity, or EVA (sometimes called a space walk), the astronaut has about 7 hours, plus or minus a half-hour, to complete his or her work. This time is dependent on the rate the astronaut is using the available air and water from the backpack. The harder the astronaut works the faster the air and water are used. The backpack also has a 30-minute emergency air supply that can be used to safely get the astronaut back inside the orbiter.

20. Is ILC responsible for the entire Shuttle space suit?

No. Actually Hamilton Standard of Windsor Locks, Connecticut is NASA's "prime contractor. " They are responsible for the Extravehicular Mobility Unit, or EMU (the entire space suit). Hamilton Standard provides the Portable Life Support System, the Display and Control Module, and several other components. As a subcontractor to Hamilton Standard, ILC furnishes the Space Suit Assembly which includes a Liquid Cooling and Ventilation Garment, Boots, Lower Torso, Arms, Gloves, and the Thermal Micrometeoroid Garment (TMG) that covers the entire EMU.

21. What do you think you will make in the year 2000?

We at ILC hope to still be making space suits for Space Shuttle and Space Station astronauts. We also hope to be designing and developing new suits for future astronauts going to the moon, Mars, and beyond!

THE NEW APOLLO AND SKYLAB SPACE SUITS

ILC Industries, Inc., through its Dover/Frederica Division in Delaware, Is the prime NASA contractor for design, development, and manufacture of space suits for the Apollo and Skylab Programs. These suits afford the necessary safety, comfort and mobility to the astronauts during Apollo and Skylab missions.

Space suit development is a tribute to American Industry. For Instance, Du Pont, the world's largest chemical corporation, developed materials used in 20 of the 21 layers In the ILC Industries space suit. None of these materials were developed with the moon in mind. Some were new materials, like "Kapton" film. Others, such as nylon, were discovered more than thirty years ago by scientists who had no Idea of the distance the results of their research would travel some day. Achievements In science are often put to use in unexpected places. In the case of the space suit, materials which were developed for use on earth ultimately found a place on the moon.

The space suit is an air-tight anthropomorphic structure called the Pressure Garment Assembly or PGA. In the space suit, the astronaut is protected from the extreme range of temperatures, the near vacuum of space and the micrometeoroid flux density that might be encountered in space or on the moon's surface. Without this protection, a man could not live, and would die within seconds after being exposed to such hostile environments.

There are two basic configurations of the suit used to support Apollo Missions: an Intravehicular (IV) configuration designated as the CMP A7LB PGA, and an Extravehicular (EV) configuration identified as the EV A7LB PGA. The CMP A7LB pressure garment configuration is worn by the Command Module Pilot. The EV A&B configuration is worn by the Crew Commander and the Lunar Module Pilot. A slightly modified version of the EV A7LB PGA is planned for use during Skylab missions.

The complete Extravehicular Mobility Units used for Apollo and Skylab missions are shown on the front cover. The Extravehicular Mobility Unit or EMU configured for Skylab missions, shown in the foreground, is the most current space suit configuration for use in space programs.

The pressure garment assembly interfaces with the spacecraft environmental control system, or the Apollo Portable Life Support System (PLSS) or the Skylab Astronaut Life Support Assembly (ALSA). The pressure garment is operational at differential pressures of 3.70 to 3.90 pounds per square Inch; In temperatures of -290 to +310 degrees Fahrenheit for Apollo missions or -180 to +277 degrees Fahrenheit for Skylab missions; and In micrometeoroid flux densities normally expected within the lunar orbit perimeter about the earth or a 300,000 mile orbit. The pressure garment permits low torque body movements for operating spacecraft controls and specially designed devices required for space exploration or traversing the lunar surface.

When pressurized, the differential pressures Impose stress or tension on the suit wall. The "soft" suit becomes very rigid or stiff, and almost impossible to bend except in those areas where specially designed joints are provided to accommodate normal body flexure. An example of this stiffness: inflate a large cylindrical balloon or the inner tube of a tire, the balloon or tube will become very stilt and almost Impossible to twist or bend. Without these specially developed joints for the space suit, It would be virtually impossible for the astronaut to do useful work on the moon's surface. These special joints are installed Into the CMP A7LB suit at the knees, wrist, shoulders, elbows, ankles, and thighs. The EV

A7LB suit was further modified to include special joints at the neck and waist to allow bending movements in those areas. This added suit flexibility permits the astronaut to conserve his energy, reduce fatigue and to work for longer periods on the lunar surface. Normal body movements in the suit cause the suit joints to bend. The force required to flex these joints is applied against the inner suit wall or gas retaining layer. To preclude direct wear on the gas-retaining layer, the suit is fitted with an inner scuff layer of nylon fabric.

the zippers in the EV A7LB suit was changed from that employed in the CMP model to accommodate the new neck and waist joints. The entrance zippers can be operated by the crewman if required, but zipper actuation is normally done with the assistance of a fellow crewmember.

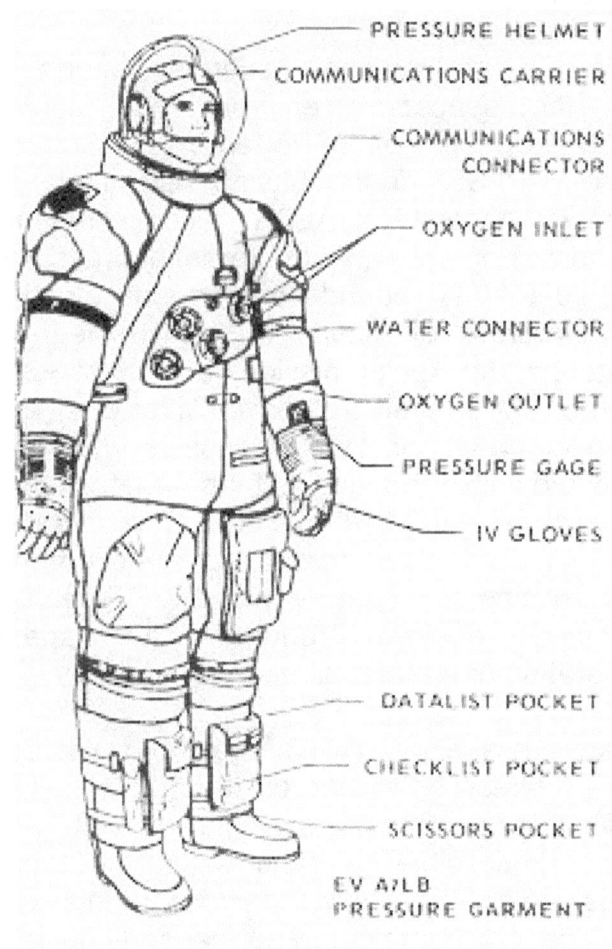

EV A7LB PRESSURE GARMENT

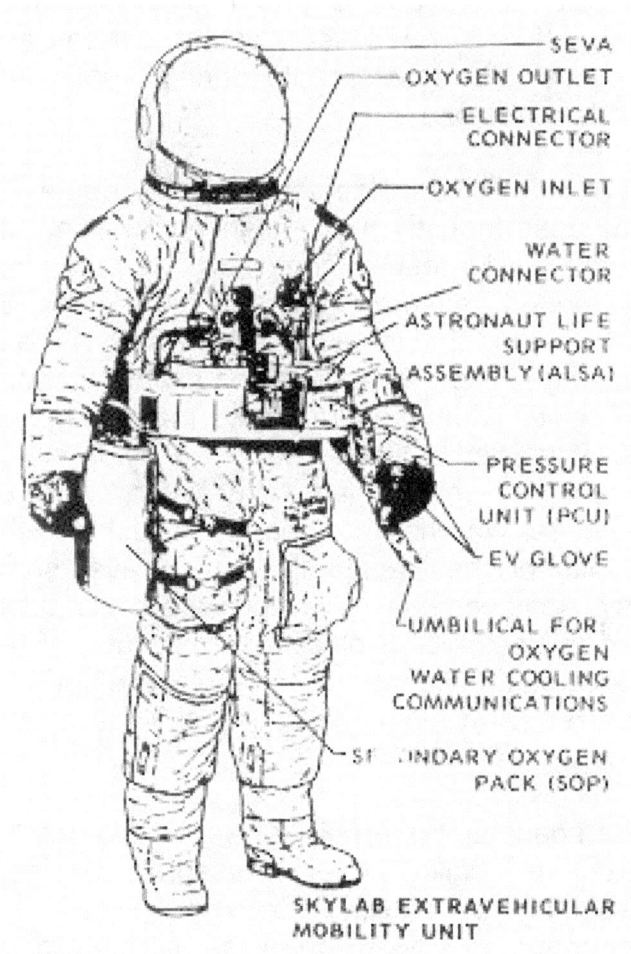

SKYLAB EXTRAVEHICULAR MOBILITY UNIT

Entrance Into the suit is made through restraint and pressure- sealing zippers. The entrance opening used In the CMP A7LB assembly extends down the center of the back, from the neck area to the front crotch area. The EV AMB suit employs entrance zippers that extend from the let t side of the waist, around the back to the right side of the waist, and diagonally up to the right chest area of the suit. Routing of

There are two protective envelopes employed in the space suit: an Inner pressurizable envelope, and an outer thermal and micrometeoroid protective envelope. The inner pressurizable envelope Is called the Torso and Limb Suit Assembly (TLSA); this assembly Interfaces with a detachable helmet, and a pair of removable gloves. The outer envelope used for thermal and micrometeoroid protection Includes an Integrated Thermal Micrometeoroid Garment (ITMG), a Lunar Extravehicular Visor Assembly (LEVA) or Skylab Extravehicular Visor Assembly (SEVA), and a pair of lunar boots that are used for Apollo Lunar missions only.

The torso and limb suit consists of an inner comfort liner, a rubber-coated nylon bladder, and an outer nylon restraint structure with the exception of the shoulder, elbow, wrist, thigh and knee joints. These joints are single wall, integrated restraint and bladder, bellows-like structures.

The Thermal Micrometeoroid Garment (ITMG) is composed of an inner layer of rubber-coated nylon, alternate layers of aluminized material separated by a low-heat-conducting spacer fabric, and an outer layer of fire and abrasion resistant material. This thermal cross section employs the same insulation principle as the "Thermos" bottle when the suit is exposed to the near vacuum of space.

The Lunar Extravehicular Visor Assembly (LEVA) or Skylab Extravehicular Visor Assembly (SEVA) includes a shell assembly that fits over the helmet, and that clamps around its base. Two visors, two side-eyeshades, and a center eyeshade are supported by the shell. The outer sun visor employs a gold coating that reflects solar heat and light from its surface. The inner protective visor is transparent, although it includes an inner coating that retains heat being emitted from the face. The protective visor is used without the sun visor during operations in shadow areas where visibility through the dark sun visor would not be adequate. The visors and eyeshades are adjustable and can be moved to positions selected by the crewman for his comfort and safety.

The Lunar Boots are slip-on assemblies that include a cross section of materials similar to those in the ITMG. There are additional layers of materials used in the boot sole as necessary to reduce the transfer of heat from the lunar surface to the foot. Metal-woven fabric or "Chromel-R" forms the outer shell of the boots to resist high lunar surface temperatures and surface abrasion. The outer structure of the boots in the sole consists of silicone rubber that is sewn to the outer metal fabric shell and affords improved wear and thermal protection to the boots.

For each PGA, there are two pairs of gloves used to support Apollo and Skylab missions: Intravehicular (IV) Gloves and Extravehicular (EV) Gloves. The IV Glove is a single-wall restraint and bladder structure formed to fit the crewman's hand. For scuff protection and added structural support, an outer gauntlet and palm restraint system is fitted over the glove. The palm restraint affords improved hand dexterity for operating spacecraft controls and special devices. The EV Glove includes an IV Glove that is fitted with an outer thermal glove that employs a similar cross section to that of the ITMG. For abrasion and thermal protection, the outer shell is constructed of metal-woven fabric, and the fingertips are fitted with silicone rubber caps. The outer thermal glove extends well back over the IV glove-TLSA juncture.

Gaseous oxygen is circulated through the suit by the PLSS (backpack) or ALSA, or the spacecraft environmental control system for respiration, pressurization, and ventilation purposes. The oxygen is directed to the helmet from inlet gas connectors on the suit, down over the body, to the arm and leg extremities, and then is directed through ducts to the exhaust gas connectors. The impurities are removed from the gas stream as it passes through the spacecraft environmental control system or portable life support system, and then is recirculated through the suit.

The ventilation system removes body heat from within the suit during IV operations, or when free space EV activities are performed remote from the spacecraft. During lunar surface excursions the metabolic heat generated by the body exceeds the capability of the ventilation system, so a liquid cooling system is employed which removes the major portion of body heat from within the PGA;

thereby reducing fatigue as a result of body dehydration through perspiration. The Liquid Cooling Garment (LCG) consists of a network of polyvinyl tubing that is supported by spandex fabric. The garment is worn next to the skin and covers the entire body exclusive of the head and hands. A liquid coolant or water is circulated through the tubing from the portable life support system. In the suit, heat is transferred from the body to the liquid through the tubing wall, and in the portable life support system, the heat is removed from the liquid before it is re-circulated back to the LCG.

Providing the spacesuit for the Apollo & Skylab Programs is just one part that ILC plays in the role of protecting man from hazardous environments. ILC Industries, Inc. is proud of its role in the space program and vigorously supports the ever-expanding field of aerospace technology. We have over twenty years of experience in research, development, and manufacture of air-inflated assemblies, pressure vessels and life support systems. This experience has provided a sound base for our continuing research, development and design of aerospace life support equipment, and has given ILC its place as the leader in this field.

Although the Apollo and Skylab programs represent a tremendous challenge and a great step forward in aerospace technology, we at ILC Industries see them as the beginning rather than the end of a long line of successful efforts. We plan to utilize our experience and knowledge gained on these programs to advance the state-of-the-art in other products for both government and Industry. The company has already utilized materials applications developed for Apollo and Skylab equipment in other products, which are being manufactured. In this manner, the results of technology gained on space programs are passed along to the consumer and the public. The walk on the moon is truly a step into the future.

On July 20, 1969, as the world watched in awe when Neil Armstrong made his "One Small Step For A Man" onto the lunar surface, a small engineering company located in Dover, Delaware was beaming with pride. That company was ILC Dover, and the pride felt at ILC was surely justified; they designed and manufactured the space suit astronaut Armstrong wore while making American history on the moon. That same pride was sustained throughout all fifteen Skylab and Apollo/Soyuz Test Project (ASTP) mission, during which ILC produced space suits performed flawlessly. It is of little wonder then, that in 1977 ILC Dover, as part of the Hamilton Standard team, was selected by NASA as Space Suit Assembly (SSA) contractor for the Space Shuttle program.

The MC Shuttle space suit is a pressure retention structure that, together with a life support system provides a life-sustaining environment, which protects the astronaut against the hazards of space. Such hazards include a vacuum environment, temperature extremes of -180 to +277 degrees Fahrenheit, and the impact of micrometeoroids and orbital debris.

Unlike the space suits used in the Apollo or Skylab Programs, where the entire suit was custom manufactured for a specific astronaut, the Shuttle suit is comprised of separate components which can be assembled to make space suits to fit almost anyone *(male & female)*. Several sizes of each component are manufactured and placed on the shelf for future use. When needed, the components are selected from the shelf *(depending on the astronaut's size)* and assembled into a complete space suit. The SSA and the Life Support System (LSS), when combined, become the Extravehicular Mobility Unit, or EMU. The EMU is used for all Shuttle program extravehicular space activities.

The SSA is designed and has been tested for an eight-year operational life. The design permits low torque body movements required for performance of tasks in space.

When pressurized, the "soft" material portion of the suit becomes very rigid and nearly impossible to bend except where specially designed joints are provided. Such is the case when you inflate the inner tube of an automobile fire.

The tube becomes very stiff and is difficult to twist or bend. Without these joints it would be virtually impossible for the astronaut to do useful work. These special joints are located at the knees, wrists, shoulders, elbows, ankles, thighs and waist of the SSA. Normal body movements by the astronaut cause the suit joints to bend. This flexibility permits the astronaut to conserve energy, reduce fatigue and to work for long periods of time.

A typical cross-section of the SSA is 11 layers deep, consisting of the liquid Cooling & Ventilation Garment (LCVG) (2 *layers*); pressure garment (2 *layers*); and the Thermal Micrometeoroid Garment (TMG) (7 *layers*). Simply stated, the LCVG maintains astronaut comfort, the pressure garment provides containment of the breathing air, and the TMG protects against the micrometeoroids which hit the suit, and insulates the astronaut from the extreme temperatures of space.

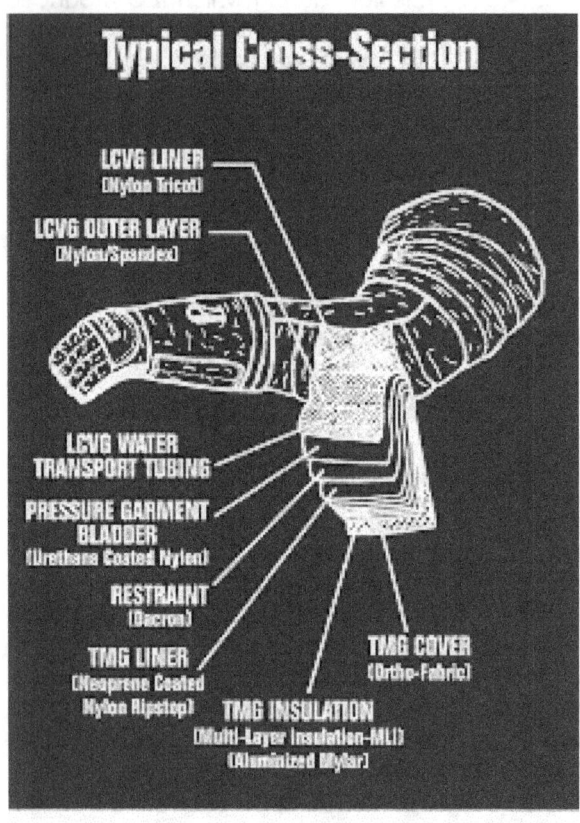

Hard Upper Torso Assembly (HIT)

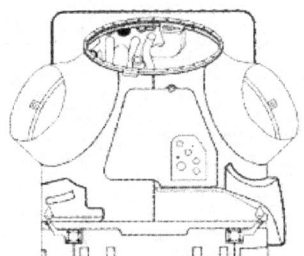

The Hard Upper Torso is a vest-like rigid fiberglass shell which incorporates provisions for Arm, LTA and Helmet attachment. A Water Line and Vent Tube Assembly is fastened to the shell interior and interfaces with he LCVG and the Life Support System (LSS). The main portion of the LSS, containing water and oxygen storage and circulation provisions, mounts on the back of the HUT, while the LSS controls mount on the front within easy reach of the astronaut.

Arm Assembly

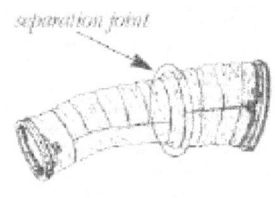

The Arm interfaces with the HUT by a ring that retains the Arm Scye Bearing in the HUT opening. The upper and lower arm joints are separated by an arm bearing, which allows lower arm rotation, the lower arm also provides for sizing adjustments and for quick connect/disconnect of the glove via a wrist disconnect.

Maximum Absorbency Garment (MAG) .

The Maximum Absorbency Garment is worn under the LCVG and provides for hygienic collection, storage, and eventual transfer of astronaut urine and feces discharged during extravehicular activities.

Communications Carrier Assembly (CCA)

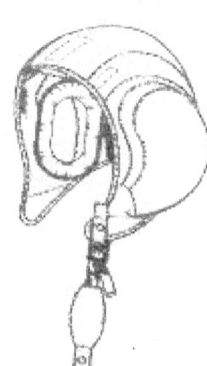

The Communications Carrier is a skull cap that interfaces with the Electrical Harness Assembly. It contains a microphone and earphones for voice communications. The skull cap is made of teflon and nylon/lycra fabrics.

Helmet Assembly

The Helmet Assembly consists of a transparent Shell, Neck Ring, Vent Pad, Purge Valve, and an adjustable Valsalva device. The Helmet is secured to the HUT and provides an unobstructed field of vision. Optical clarity of the transparent shell is made possible by the use of rugged, impact resistant polycarbonate material. A vent assembly, bonded to the inside rear of the polycarbonate shell, serves to diffuse the incoming gas over the astronaut's face.

Lower Torso Assembly

separation joints

The Lower Torso Assembly consists of an integrated Body Seal Closure, Waist, Waist Bearing, Leg, Thigh, Knee and Ankle joints, plus Boots. The LTA encloses the lower body and interfaces with the HUT via the body seal closure. The flexible waist section and waist bearing afford the astronaut a large degree of movement about the waist, e.g. bending and hip rotation.

Glove Assembly

The Glove is made up of a restraint and bladder encased in a TMG. The gloves protect the astronaut's wrists and hands and are attached to the arms at the wrist disconnects. The gloves incorporate a rotary bearing to allow wrist rotation, a wrist joint to provide flexion/extension, fabric joints for thumbs and fingers, plus a hot pad for protection of the hand from extreme hot and cold extravehicular conditions. The glove includes fingertip heaters that are controlled by the astronaut.

Extravehicular Visor Assembly (EVVA)

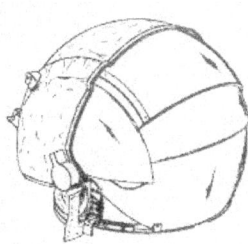

The Extravehicular Visor Assembly is a light-and-heat-attenuating shell which fits over the Helmet Assembly. It is designed to provide protection against micrometeoroid activity and accidental impact damage, plus protect the crewmember from solar radiation. A special coating gives the sun visor optical characteristics similar to those of a two-way mirror; it reflects solar heat and light, yet permits the astronaut to see. Adjustable eyeshades may be pulled down over the visor to provide further protection against sunlight and glare.

Liquid Cooling & Ventilation Garment (LCVG)

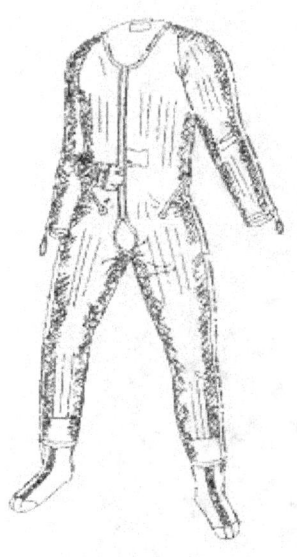

The liquid Cooling & Ventilation Garment is a close-fitting undergarment covering the body torso and limbs. It incorporates a network of fine tubing that is maintained in close contact with the astronaut's skin by an outer layer of stretchable open fabric. The space suit is so well insulated that normal body heat maintains warmth, except for occasional cold hands, even on the cold, dark side of the spacecraft. However, cooling is required, therefore, water is circulated through the LCVG tubing to remove excess body heat. Water flows through the various inlet and return tubes and must be uninterrupted in order for the garment to be effective. The LCVG also uses ventilation ducting to return vent flow from the body extremities to the EMU Life Support System (LSS).

Insuit Drink Bag (IDB)

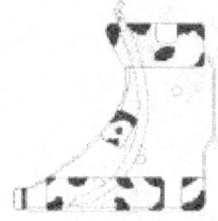

The IDB is a sealed bag that comes in two sizes, holding 21 oz. and 32 oz. of potable *(drinking)* water. The bag is secured by velcro to the inside front of the HUT. Water is readily accessible to the astronaut through a mouthpiece located at the top of the bag.

Since 1947, ILC has been active in the development of products for both government and industry. Then as today, most A C products are comprised of softgoods materials - our primary area of expertise. Today, ILC's products fall into six primary groups:

- Space Suits and Equipment
- Environmental Protection Products
- Individual and Collective Protection Equipment
- High Technology Inflatables
- Camouflage, Concealment, and Deception
- Human Engineered Composites

Space Suit Assembly

Following are brief descriptions and illustrations of the units that comprise the Space Suit Assembly:

1. Communications Carrier Assembly (CCA)
2. Hard Upper Torso Assembly (HUT)
3. Arm Assembly
4. Maximum Absorbency Garment (MAG)
5. Helmet Assembly
6. Lower Torso Assembly (LTA)
7. Glove Assembly
8. Extravehicular Visor Assembly (EVVA)
9. Liquid Cooling & Ventilation Garment (LCVG)
10. Insuit Drink Bag (IDB)

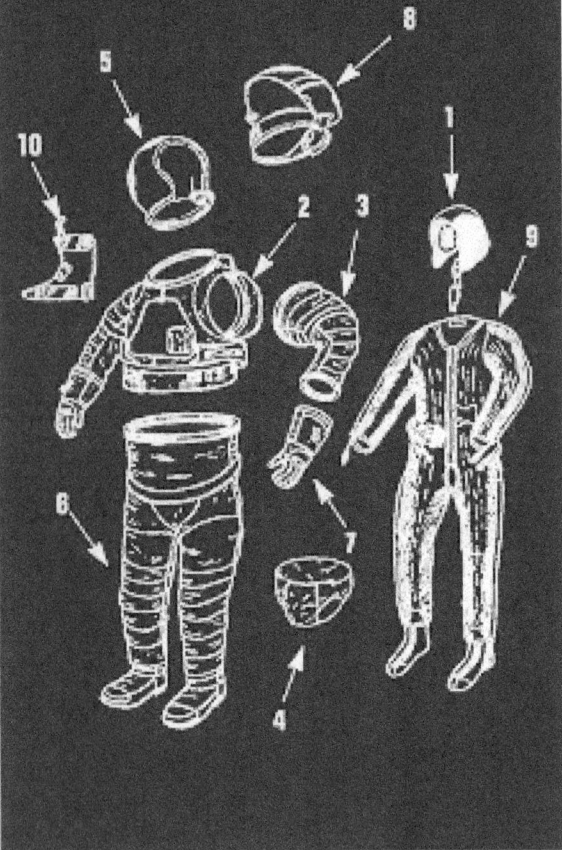

NASA

National Aeronautics and
Space Administration

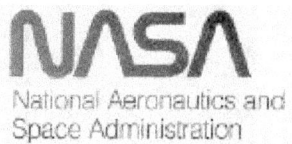

Information Summaries

PMS-033 (JSC)
July 1989

Wardrobe for Space

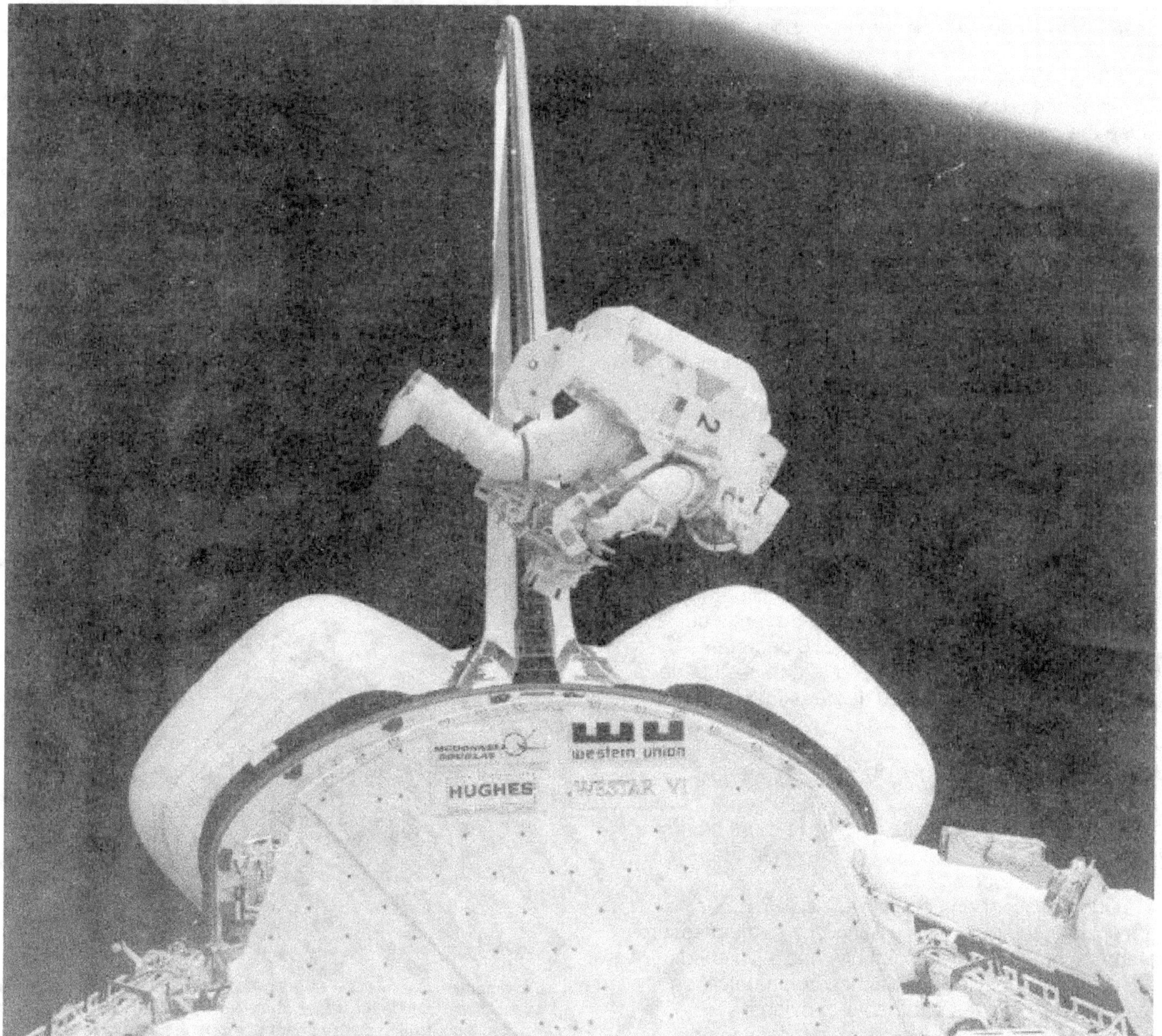

An astronaut working in the space shuffle orbiter cargo bay wears a spacesuit and a jet-powered, hand-controlled maneuvering unit. With this protection and equipment, the astronaut becomes an individual spacecraft able to move about and perform tasks in the microgravity of space.

WHY WEAR A SPACESUIT?

To explore and work in space, human beings must take their environment with them because there is no atmospheric pressure and no oxygen to sustain life. Inside the spacecraft, the atmosphere can be controlled so that special clothing isn't needed, but when outside, humans need the protection of a spacesuit.

Earth's atmosphere is 20 percent oxygen and 80 percent nitrogen from sea level to about 75 miles up, where space begins. At 18,000 feet, the atmosphere is half as dense as it is on the ground, and at altitudes above 40,000 feet, air is so thin and the amount of oxygen so small that pressure oxygen masks no longer do the job. Above the 63,000-foot threshold, humans must wear spacesuits that supply oxygen for breathing and that maintain a pressure around the body to keep body fluids in the liquid state. At this altitude the total air pressure is no longer sufficient to keep body fluids from boiling.

Spacesuits for the space shuttle era are pressurized at 4.3 pounds per square inch (psi), but because the gas in the suit is 100 percent oxygen instead of 20 percent, the person in a spacesuit actually has more oxygen to breathe than is available at an altitude of 10,000 feet or even at sea level without the spacesuit. Before leaving the space shuttle to perform tasks in space, an astronaut has to spend several hours breathing pure oxygen before proceeding into space. This procedure is necessary to remove nitrogen dissolved in body fluids and thereby to prevent its release as gas bubbles when pressure is reduced; a condition commonly called "the bends."

Spacesuits designed for the space station era will be pressurized to 8.3 psi; therefore, the pre-breathing period will be shortened or diminished.

The spacesuit also shields the astronaut from deadly hazards. Besides providing protection from bombardment by micrometeoroids, the spacesuit insulates the wearer from the temperature extremes of space. Without the Earth's atmosphere to filter the sunlight, the side of the suit facing the Sun may be heated to a temperature as high as 250 degrees Fahrenheit; the other side, exposed to darkness of deep space, may get as cold as -250 degrees Fahrenheit.

WARDROBE FOR THE SPACE SHUTTLE ERA

Astronauts of the space shuttle era have more than one wardrobe for space flight and what they wear depends on the job they are doing.

During ascent and entry, each crewmember wears special equipment consisting of a partialpressure suit, a parachute harness assembly, and a parachute pack. The suit, consisting of helmet, communication assembly, torso, gloves and boots, provides counterpressure and anti-exposure functions in an emergency situation in which the crew must parachute

STS-26 crewmembers leave the Operations and Checkout Building at Kennedy Space Center, Florida, heading for the launch pad and lift-off. They are wearing partial pressure suits developed for the launch and entry phases of space flight.

from the orbiter, The suit has inflatable bladders that fill it with oxygen from the orbiter. These bladders inflate automatically at reduced cabin pressure. They also can be manually inflated during entry to prevent the crewmember

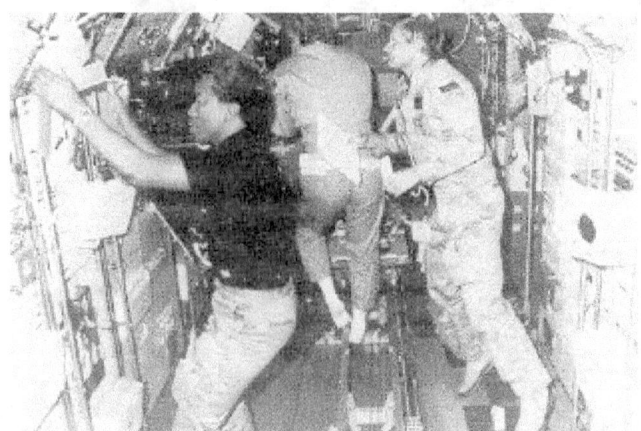

Crewmembers of mission STS 61-A perform experiments in the spacelab module in the space shuttle orbiter cargo bay. Astronauts work in a controlled environment of moderate temperatures and unpolluted air at sea-level atmosphere.

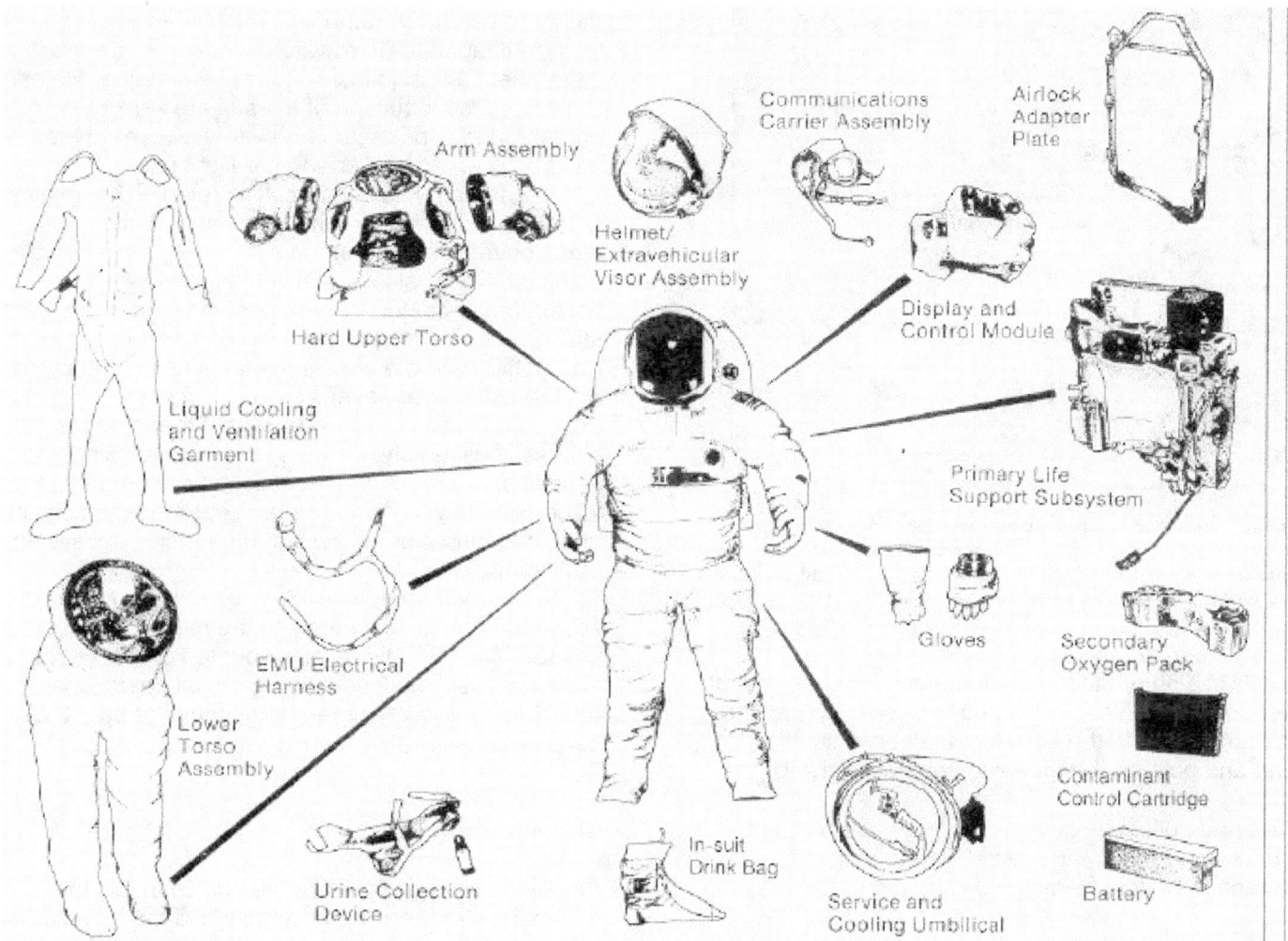

Space shuttle EMU components

from blacking out. Without the suit pressing on the abdomen and the legs, the blood would pool in the lower part of the body and cause a person to black out as the spacecraft returns from microgravity to Earth's gravity. The partial-pressure suit and equipment will support a crewmember for a 24-hour period in a liferaft in case of an egress over water.

Working Inside the Space Shuttle

During orbit, astronauts work inside the space shuttle in shirtsleeve comfort. Prior to a mission, crewmembers are outfitted from a selection of clothing including flight suits, trousers, lined zipper jackets, knit shirts, sleep shorts, soft slippers, and underwear. The materials of every component of the clothing are flame retardant. Covering the exterior of the garments are closable pockets for storing such items as pens, pencils, data books, sunglasses, a multipurpose Swiss army pocketknife, and scissors.

Working Outside the Space Shuttle

To work in the open cargo bay of the space shuttle or in space, astronauts wear the shuttle extravehicular mobility unit (EMU) spacesuit, which was developed to be More durable and more flexible than previous spacesuits were. The suit is modular in design, with many interchangeable parts. The upper torso, lower torso, arms, and gloves are manufactured in different sizes and can be assembled for each mission in combinations needed to fit men and women astronauts. This design is cost-effective because the suits are reusable and not custom fitted as were spacesuits used in previous NASA manned space flight programs.

Suiting up

The EMU comprises the spacesuit assembly, the primary life support system (PLSS), the display and control module, and several other crew items designed for spacewalks and emergency life support. The EMU accommodates a variety of interchangeable systems that interconnect easily and

Astronaut Sherwood C. Spring wears the EMU while checking joints on a tower extending from the Space Shuttle Atlantis cargo bay during mission STS 61-B. He is standing on the end of the arm of the remote manipulator system connected to the space shuttle.

securely in single-handed operation for either normal or emergency use. When preparing to work in space, the astronaut goes into the airlock of the space shuttle orbiter and puts on the following parts of the EMU:

- A urine-collection device that receives and stores urine for transfer later to the orbiter waste management system.

- A liquid cooling and ventilation garment, a one-piece mesh suit made of spandex, zippered for front entry, and weighing 6.5 pounds dry. The garment has water-cooling tubes running through it to keep the wearer comfortable during active work periods.

- An in-suit drink bag containing 21 ounces of potable water, the "Snoopy Cap," or communications carrier assembly, with headphones and microphones for two-way communications and caution-and-warning tones, and a biomedical instrumentation subsystem.

To put on the spacesuit, the astronaut first dons the lower torso assembly and then rises into the top section of the two-piece EMU spacesuit hanging on the wall of the airlock. The upper torso of the spacesuit is a hard-shell fiberglass structure that contains the primary life support system and the display control module. Connections between the two parts must be aligned to enable circulation of water and gas into the liquid cooling ventilation garment and return. Then, the gloves are added and last to be donned is the extravehicular visor and helmet assembly, which provides protection from micrometeoroids and from solar ultraviolet and infrared radiation. Bearings in the shoulder, arm, wrist, and waist joints allow the crewmember freedom of movement. Bending, leaning, and twisting motions of the torso can all be done with relative ease.

All fabric-to-hardware connections are made with either mechanical joints or adhesive bonding. Materials used in the construction of the suit are selected to prevent fungus or bacteria growth; however, the suit Must be cleaned and dried after flight use.

The entire suit assembly is rated with a minimum 8-year life expectancy. The nominal operating atmospheric pressure in the suit is 4.3 psid. The suit comprises several layers including a polyurethane-coated nylon pressure bladder, a polyester structural restraint layer with folded and pleated joints (for mobility), and a woven Kevlar, Teflon, and Dacron anti-abrasion outer layer.

The maximum total weight of the largest size spacesuit assembly, including the liquid cooling and ventilation garment, urine collection device, helmet and visor assembly, communications carrier assembly, insuit drink bag, and biomedical instrumentation subsystem, is 107 pounds,

The astronaut is ready to go to work in space and secures the necessary tools to the mini-workstation of the suit. The EMU lights are mounted on the helmet and are a necessity because during orbital operations approximately every other 45 minutes are spent in darkness.

Communications

An electrical harness inside the suit connects the communications carrier assembly and the biomedical instrumentation equipment to the hard upper torso where internal connections are routed to the extravehicular communicator by means of a passthrough.

The extravehicular communicator attaches to the upper portion of the life support system at the back of the hard upper torso, The controls are located on the display and control module mounted on the chest at the front of the upper torso. The extravehicular communicator provides radio communication between the suited crewmember and the orbiter. In addition, electrocardiographic (EKG) information is telemetered through the extravehicular communicator to the orbiter and to flight surgeons in the Mission Control Center at Houston, Texas.

The radios for spacewalk communications have two single UHF channel transmitters, three singlechannel receivers, and a switching mechanism. These backpack radios have a "low profile" antenna - a footlong rectangular block fitted to the top of the PLSS. The radios weigh 8.7 pounds and are 12 inches long, 4.3 inches high, and 3.5 inches wide.

Primary life support system

The PLSS consists of a backpack unit permanently mounted to the hard upper torso of the suit and a

Astronaut Bruce McCandless II called the MMU "a great flying machine." McCandless was the first astronaut to use the MMU for a spacewalk on February 7, 1984, during space shuffle mission STS 41-B.

Control-and-display unit mounted on the suit chest. The backpack unit supplies oxygen for breathing, suit pressurization, and ventilation. The unit also cools and circulates water used in the liquid cooling ventilation garment, controls ventilation gas temperature, absorbs carbon dioxide, and removes odors from the suit atmosphere. The secondary oxygen pack attaches to the bottom of the PLSS and supplies oxygen if the primary oxygen fails. The control-and-display unit allows the crewmember to control and monitor the PLSS, the secondary oxygen pack, and, when attached, the manned maneuvering unit.

Maneuvering in space

The manned maneuvering unit (MMU) is a one-man, nitrogen-propelled backpack that latches to the EMU spacesuit's PLSS. Using rotational and translational hand controllers, the crewmember can fly with precision in or around the orbiter cargo bay or to nearby freeflying payloads or structures, and can reach many otherwise inaccessible areas outside the orbiter. Astronauts wearing MMU's have deployed, serviced, repaired, and retrieved satellite payloads.

The MMU propellant - noncontaminating gaseous nitrogen stored under high pressure - can be recharged from the orbiter, The reliability of the unit is guaranteed with a dual parallel system rather than a backup redundant system. In the event of a failure in one parallel system, the system would be shut down and the remaining system would be used to return the MMU to the orbiter cargo bay. The MMU, which weighs 310 pounds, includes a 35-mm still photo camera that is operated by the astronaut while working in space.

SPACESUITS FOR SPACE STATION AND BEYOND

When Edward White opened the Gemini IV hatch in 1965 and became the first American to step into the vacuum of space, life-giving oxygen was fed to his spacesuit by a 25-foot umbilical to a chest-mounted pressure regulator and ventilation assembly.

Exploring the Moon, however, required tore independence from the spacecraft. The backpack portable life support system, a self-contained supply of breathing and pressurizing oxygen, filters for removing carbon dioxide, and cooling water, gave Apollo crewmen the independence they needed.

The PLSS supplied oxygen at 3.5 to 4 psi of pressure while circulating cooling water through the liquid-cooling garment worn under the spacesuit. Lithium hydroxide filters removed carbon dioxide from the crewman's exhaled breath, and charcoal and Orlon

The Mark III suit is modeled at Johnson Space Center, where it is being evaluated for use in the space station era

filters sifted out odors and foreign particles from the breathing oxygen. Metabolic heat was transferred from the cooling water loops to space through a water evaporator system in the PLSS.

Mounted atop the PLSS was an emergency 30-minute supply of oxygen and communications equipment for talking with fellow crewmen on the lunar surface and with flight controllers in Mission Control Center in Houston. Additionally, the communications systems relayed back to Earth biomedical data on the crewmen.

The Apollo lunar surface spacesuit and PLSS weighed 180 pounds on Earth, but only 30 pounds on the Moon because of the difference in gravity.

With the advent of the space station program, work outside the spacecraft will be expanded to provide numerous unique on-orbit service capabilities not fully achieved in previous space program operations. Currently, scientists and engineers at two NASA facilities are working on a new generation of spacesuits for use during activity in space.

The Mark III suit, a combination of hard and soft elements, is being developed at the NASA Lyndon B Johnson Space Center (JSC) in Houston, Texas. The AX-5, a hard, all metal suit, is being developed by the NASA Ames Research Center (ARC) in California. Both suits share common design goals. For example, they must be easy to get into and out of, must be comfortable to wear, and must allow adequate mobility and range of motion for the jobs to be performed. Both are designed to be altered to fit different size astronauts

For use on space station, the suit must also be easily maintained and provide necessary protection from radiation, micrometeoroids, and manmade debris. In addition to these requirements, both the JSC Mark III and the ARC AX-5 suit have been designed to operate at a pressure of 8.3 psi. Current space shuttle spacesuits operate at 4.3 psi and require a time-consuming pre-breathing operation prior to the beginning of any spacewalk.

Pre-breathing allows the astronaut's body to adapt to the difference in pressure between the spacecraft cabin and the suit. By operating at a higher pressure, which more closely matches that of the space station, the new suit would greatly reduce or even eliminate the need for pre-breathing. Astronauts in the space station will be able to prepare for outside activity in much less time.

HISTORY OF WARDROBES FOR SPACE

The Mercury spacesuit was a modified version of a U.S. Navy high altitude jet aircraft pressure suit. It consisted of an inner layer of Neoprene-coated nylon fabric and a restraint outer layer of aluminized nylon. Joint mobility at the elbow and knees was provided by simple fabric break lines sewn into the suit; but even with these break lines, it was difficult for a pilot to bend his arms or legs against the force of a pressurized suit. As an elbow or knee joint was bent, the suit joints folded in on themselves reducing suit internal volume and increasing pressure.

The first seven astronauts selected by NASA in 1959 wear the spacesuits developed for Project Mercury.

Astronaut Edward H. White 11 was the first U.S. crewman to float in the zero gravity of space. His spacewalk occurred during the flight of Gemini 4.

Apollo 17 scientist-astronaut Harrison H. Schmitt wears a protective spacesuit and helmet while collecting lunar samples at the Taurus-Littrow landing site in 1972. His backpack is the portable life support system.

The Mercury suit was worn "soft" or unpressurized and served only as a backup for possible spacecraft cabin pressure loss - an event that never happened. Limited pressurized mobility would have been a minor inconvenience in the small Mercury spacecraft cabin.

Spacesuit designers followed the U.S. Air Force approach toward greater suit mobility when they began to develop the spacesuit for the two-man Gemini spacecraft. Instead of the fabric-type joints used in the Mercury suit, the Gemini spacesuit had a combination of a pressure bladder and a link-net restraint layer that made the whole suit flexible when pressurized.

The gas-tight, man-shaped pressure bladder was made of Neoprene-coated nylon and covered by loadbearing link-net woven from Dacron and Teflon cords. The net layer, being slightly smaller than the pressure bladder, reduced the stiffness of the suit when pressurized and served as a sort of structural shell, much like a tire contained the pressure load of the innertube in the era before tubeless tires, Improved arm and shoulder mobility resulted from the multilayer design of the Gemini suit.

Tailored for Apollo Moon Walking

Walking on the Moon's surface a quarter million miles away from Earth presented a new set of problems to spacesuit designers. Not only did the Moon explorers' spacesuits have to offer protection from jagged **rocks and** the searing heat of the lunar day, but the suits also had to be flexible enough to permit stooping and bending as Apollo crewmen gathered samples from the Moon, set up scientific data stations at each landing site, and used the lunar rover vehicle, an electric-powered dune buggy, for transportation over the surface of the Moon.

The additional hazard of micrometeoroids that constantly pelt the lunar surface from deep space was met with an outer protective layer on the Apollo spacesuit. A backpack portable life support system provided oxygen for breathing, suit pressurization, and ventilation for moonwalks lasting up to 7 hours,

Apollo spacesuit mobility was improved over earlier suits by use of bellows-like molded rubber joints at the shoulders, elbows, hips and knees. Modifications to the suit waist for Apollo 15 through 17 missions added flexibility making it easier for crewmen to sit on the lunar rover vehicle.

From the skin out, the Apollo A7LB spacesuit began with an astronaut-worn liquid-cooling garment, similar to a pair of longjohns with a network of spaghettilike tubing sewn onto the fabric. Cool water, circulating through the tubing, transferred metabolic heat from the Moon explorer's body to the backpack and thence to space.

Next came a comfort and donning improvement layer of lightweight nylon, followed by a gas-tight pressure bladder of Neoprene-coated nylon or bellows-like molded joints components, a nylon restraint layer to prevent the bladder from ballooning, a lightweight thermal superinsulation of alternating layers of thin Kapton and glass-fiber cloth, several layers of Mylar and spacer Material, and finally, protective outer layers of Tefloncoated glass-fiber Beta cloth.

Apollo space helmets were formed from highstrength polycarbonate and were attached to the spacesuit by a pressure-sealing neckring. Unlike Mercury and Gemini helmets, which were closely fitted and moved with the crewman's head, the Apollo helmet was fixed and the head was free to move within. While walking on the Moon, Apollo crewmen wore an outer visor assembly over the polycarbonate helmet to shield against eyedamaging ultraviolet radiation, and to maintain head and face thermal comfort.

Completing the Moon explorer's ensemble were lunar gloves and boots, both designed for the rigors of exploring, and the gloves for adjusting sensitive instruments.

The lunar surface gloves consisted of integral structural restraint and pressure bladders, molded from casts of the crewmen's hands, and covered by multilayered superinsulation for thermal and abrasion protection. Thumb and fingertips were molded of silicone rubber to permit a degree of sensitivity and "feel." Pressure-sealing disconnects, similar to the helmet-to-suit connection, attached the gloves to the spacesuit arms.

The lunar boot was actually an overshoe that the Apollo lunar explorer slipped on over the integral pressure boot of the spacesuit. The outer layer of the lunar boot was made from metal-woven fabric, except for the ribbed silicone rubber sole; the tongue area was made from Teflon-coated glass-fiber cloth. The boot inner layers were made from Teflon-coated glass-fiber cloth followed by 25 alternating layers of Kapton film and glass-fiber cloth to form an efficient, lightweight thermal insulation.

Wardrobe for Skylab and Apollo-Soyuz

Nine Skylab crewmen manned the Nation's first space station for a total of 171 days during 1973 and 1974. They wore simplified versions of the Apollo spacesuit while doing the historic repair of the Skylab and changing film canisters in the solar observatory cameras. Jammed solar panels and the loss of a micrometeoroid shield during the launch of the Skylab orbital workshop necessitated several spacewalks for freeing the solar panels and for erecting a substitute shield.

The spacesuit changes from Apollo to Skylab included a less expensive to manufacture and lightweight thermal micrometeoroid overgarment, elimination of the lunar boots, and a simplified and less expensive extravehicular visor assembly over the helmet. The liquidcooling garment was retained from Apollo, but umbilicals and astronaut life support assembly (ALSA) replaced backpacks for life support during spacewalks.

Apollo-type spacesuits were used again in July 1975 when American astronauts and Soviet cosmonauts rendezvoused and docked in Earth orbit in the joint Apollo-Soyuz Test Project (ASTP) flight. Because no spacewalks were planned, U.S. crewmen were equipped with modified A7LB intravehicular Apollo spacesuits fitted with a simple cover layer replacing the thermal micrometeoroid layer.

Visitors to the Johnson Space Center in Houston, Texas, can see an exhibit of spacesuits.

Suit Sizing
New space suits can be sized in space saving storage, deliveries

By Karen Schmidt

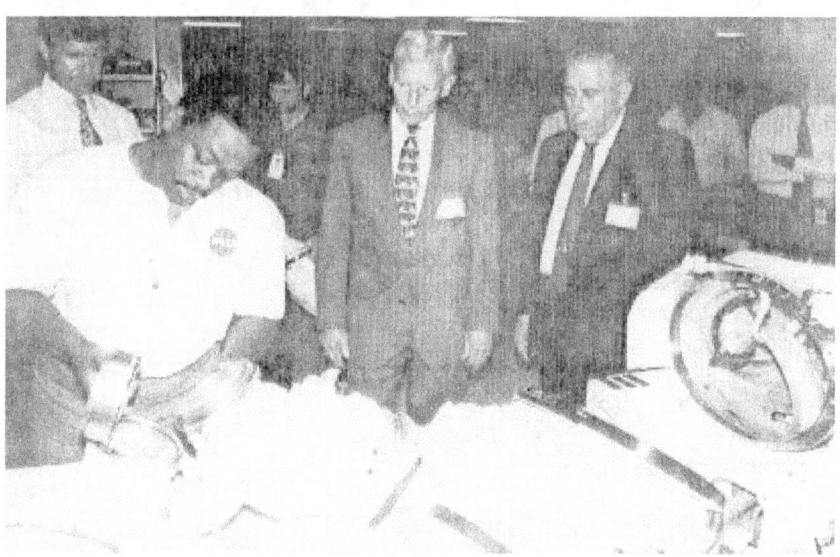

The STS-79 mission will carry into orbit a new space walking suit designed to fit more than one astronaut and save storage space, bringing tomorrow's International Space Station technology into today's missions.

Extravehicular Activity space suits are the astronauts' lifeblood when they must work outside the protected environment of a space shuttle. Equipped with life support, an astronaut can spend up to seven hours performing maintenance tasks in the shuttle's cargo bay or on the future space station. With EVA tasks expected to increase during station assembly and operation, **JSC, in cooperation with Hamilton Standard, ILC Dover, Air-Lock and** Boeing Aerospace Operations, is revamping space suits to save storage space, meet weight limits and reduce the amount of equipment required on flights to the station.

One of the first phases of the redesign was to develop a way to resize a suit faster on the ground and in orbit. Currently, ground technicians change suit sizes by lacing in different lengths of fabric inserts.

"In order to make a suit fit an astronaut, technicians must change the inserts in the arms and legs of a suit," said Ralph Anderson of the Flight Crew Equipment Management Office. "It is a long and cumbersome process that takes about 16 hours to prepare a suit for a particular astronaut."

Astronauts also are trained to change-out inserts in the suits, but the process is slow and tedious, taking up valuable on-orbit time. The new design features sizing rings in both arms and legs that can be changed out in less time.

"With the enhanced sizing rings, a suit technician can change the size of a suit in less than 20 minutes," Anderson added.

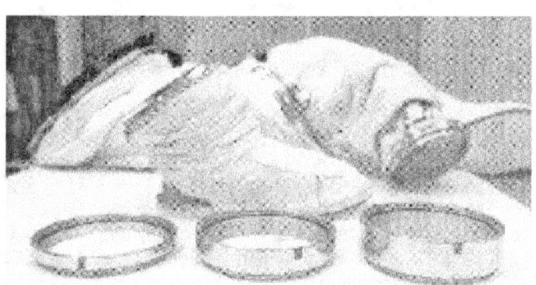

Not only can suit technicians change the size of an EVA suit, the astronauts on orbit will have the same capability.

"That's the whole idea, multiple crew members can use the same suit for space walks on the International Space Station," Anderson said.

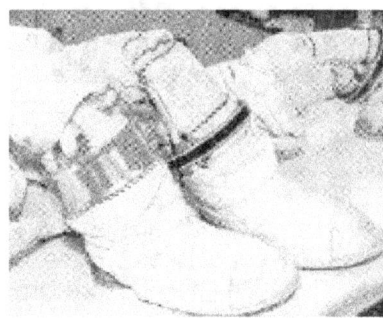

The new rings, made of aluminum, are available in 1/2 size at the arm and thigh and three different sizes for the lower leg-1/2, 1 and 1 1/2 inch. There also are four different sizes of leg segments and eight sizes of lower arm segments that the astronauts can choose from. One leg attachment that fits from thigh to ankle can be sized up to three inches-with so many combinations, a single suit can be sized to fit a number of astronauts.

"We will be able to carry a couple of suits and leave them on the space station with enough sizing components to fit different astronauts, thereby eliminating the need to carry suits for specific astronauts on every flight," said Rodney Johnson, lead for the Training Extravehicular Mobility Unit Laboratory at Boeing.

The design of the sizing rings evolved from rings used on an advance development suit. The major difference is that the new rings are threaded and twist on. Each ring has two automatic spring locks and one manual lock.

The new ring uses a pressure seal that is an adaptation of the static seal that we have been using for a decade and a half in disconnects found at the neck, gloves and waist," said Don Lacey of ILC Dover. "Adapting proven designs reduced our learning curve tremendously. The suit was designed to meet the space station mission, but we will begin to reap the benefits of this new suit right away."

In addition to the new rings, Adjustable Restraint Brackets also are being used for the first time. They allow astronauts to lengthen or shorten the arm and leg segments in smaller increments than the rings.

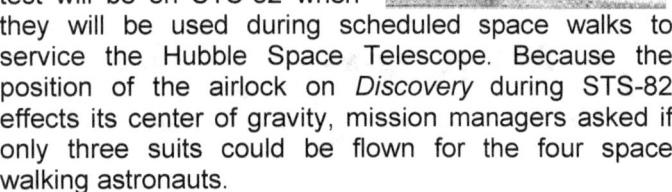

"You can lengthen either end up to one half inch," said Scott Cupples of ILC Dover, giving an astronaut a custom fit."

The suits will fly for the first time on STS-79, but the big test will be on STS-82 when they will be used during scheduled space walks to service the Hubble Space Telescope. Because the position of the airlock on *Discovery* during STS-82 effects its center of gravity, mission managers asked if only three suits could be flown for the four space walking astronauts.

Mission Operations STS82 EMU Lead Paul Boehm and back-up Dana Weigel were able to answer 'yes' because of the new sizing capabilities. Mission Specialists Joe Tanner and Steve Smith will share one suit, bringing enough sizing rings and leg attachments to custom fit the two space walkers.

"We are relying on these rings to accomplish a resize in a much shorter time," Boehm said. "It is going to be nice to have the capability to do **this on orbit,** it gives us a lot more flexibility and helps us focus on the primary objectives of the mission."

"The sizing rings add a new, much needed capability to resize the EVA suits in flight," Tanner said. "This capability allows us to carry only three suits to accommodate four EVA crew members. The rings are very easy to use, requiring only a few minutes to change arm and or leg segments to fit another crew member. The rings don't restrict your motion in the suit in any way, in fact, I can't even tell they are there. Other modifications that go along with the enhanced EMU allow crew members to make minor adjustments to arm and leg segment lengths that could previously only be made by a ground technician. The end result is a better suit with more capability and flexibility to carry us into the station era."

More redesigns are in the works. The Hard Upper Torso, or HUT, will be fitted with new quick disconnects instead of bolt on attachments that connect the Primary Life Support System. These quick disconnects are expected to work better and faster and the pivot points at the shoulders of the current HUT will be deleted to give the astronauts better mobility and make the suit more robust. The new designed HUT also will remove four possible failure points that now exist in the older model.

"If we had a space station today, this suit would be ready to fly," said Tony Wagner, spacesuit subsystem manager in the Crew and Thermal System EMU Group. "We could leave it on the station for many space walks before it would have to return for maintenance. It is certified and ready to go for EVA."

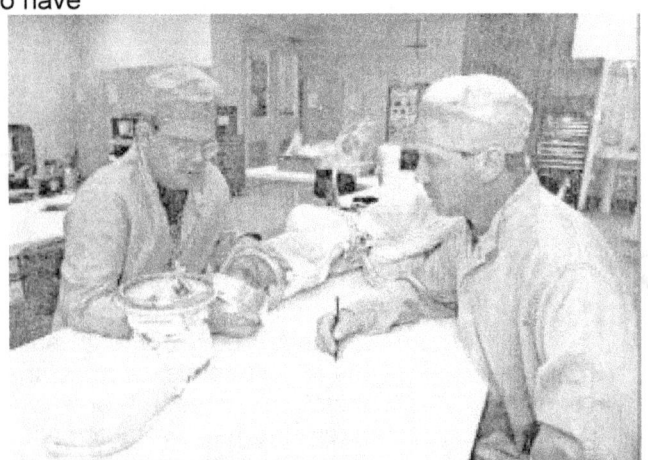

From top to bottom, left to right: 1) From left, Rodney Johnson, lead for the Training Extravehicular Mobility Unit Laboratory at Boeing Aerospace Operations, demonstrates how the new enhanced sizing rings work to NASA Deputy Administrator John Dailey and JSC Director George Abbey. 2) The new rings replace fabric sizing inserts that were hand laced into the suit by technicians, taking up to 16 hours to resize on suit. 3) Leg rings come in three different sizes assuring a custom fit for the astronauts. 4) Another new design feature is an Adjustable Restraint Bracket that gives astronauts a second length adjustment feature in the arms and legs of the space suit. 5) From left, Robert Nicholson and Ron Lindsey prepare a suit in the flight EMU laboratory for STS-79. 6) From left, Latonya Hagler and Nicholas Barnett check an arm ring assembly.

MILESTONES AND CAPSULE HISTORY
ILC Dover, Inc.

1947 International Latex Corp. divides into 4 separate divisions. One is an outgrowth of the specialty group named the Metals Division located at Pear and Mary Streets in Dover, Delaware. The Metals Division makes display racks for Playtex products such as girdles, swim caps, baby pants as well as life vests, life rafts, and ant-exposure clothing. The other three divisions become Playtex (rubber and fabric goods), Chemical Division (now known as Reichhold Chemical Co.) and the Pharmaceutical Division (which makes items such as throat lozenges).

1952 Metals Division is awarded a contract to supply the Navy and Air Force with high altitude pressure helmets. The work force at this time is about 30 people.

1945 Stanley Warner Corp. purchases the controlling interest in the International Latex Corp.

1955 Metals Division changes its name to the Special Products Division.

1956 The high altitude pressure helmet contract is expanded to include high altitude pressure suits. The fact that the Special Products Division was involved in rubber compounding and dipping methods helped resolve the problems of flexible joints in these suits (elbows, knees, and waists). Work continued on life rafts for the Army and prototype helmets for the space program.

1958 Special Products Division became known as the Industrial Products Division.

1960 Production is busy with soft-goods assembly work which includes ammunition pouches, harness straps and first aid pouches.

A facility in Frederica is used as a warehouse to support this production.

1962 The Industrial Products Division begins work on the Apollo Space Suit as a subcontractor to Hamilton Standard. The work force at this time numbers 50.

1964 The Industrial Products Division is re-named the Government and Industrial Division.

1966 Government and Industrial Division is awarded prime contract on the Apollo Lunar Space Suit. Our work force increases to 200.

1966/1967 International Latex Corporation formally splits into three separate entities: IPC (Playtex), Standard Brands (now Reichhold), and ILC Industries (formally the Government and Industrial division). In Nov. '66 ILC Industries, Inc. is incorporated in the State of Delaware as a wholly owned subsidiary of Stanley Warner Corp. On Dec. 22 1967 Stanley Warner Corp. is merged into the Glen Alden Corp.

In 1967, the company developed at its own expense, a barrier bag used to protect consumable-combustible cartridge cases for munitions. The company eventually sold 475,000 barrier bags to the US Army.

Under an Air Force contract, ILC starts the production of 384 double-wall air supported structures in the Frederica facility.

As a spin-off of the double-wall structures, ILC starts designing, fabricating and installing single-wall air-supported structures used as warehouses, field houses, and tennis court and swimming pool enclosures.

Drawing on experience gained on high altitude helmet manufacturing in the early 50's, ILC Industries, Inc. starts 8 line of riot helmets and face-shields that were used by police and fire departments in larger cities during the late 1960's. At this time ILC Industries Inc. was manufacturing motorcycle helmets along with the Vari-Shield face shield which was popular with motorcyclists and snow-mobilers alike.

1968 Our combined work force In Dover and Frederica numbers 755.

At this time, 90% of the company's revenues come from the space suit program.

We begin production of 1000 inflatable boats for the US Army.

1969 In Jan. of 1969, ILC went public and sold 225,000 shares of common stock to the public at $7.25 a share. Glen Alden retained 70% ownership of the company. (All outstanding public shares were purchased by Rapid American, when they acquired Glen Alden)

The proceeds of the sale were used to purchase Bio Medical Electronics Inc., a producer of patient monitoring equipment that went out of business in 1972.

July 20, 1969 will be remembered by many ILC employees to be one of the highlights of their lifetime. Nell Armstrong, attired in an ILC Industries Inc. designed and manufactured Apollo space suit took man's first step on the moon. An interesting fact of that mission and other lunar missions was that the space suit was the only item to be used on the lunar surface and brought back to earth, unlike their other flight equipment. Our space suits were used on the remainder of the Apollo flights and in Skylab and the Apollo Soyuz Test Project. It is worth mentioning that of the 15 missions our suits were flown on, they performed flawlessly.

1970 our work force has grown to goo. ILC Industries Inc. wins a contract to rebuild portions of the Family 11 200,000 cubic foot balloons in conjunction with the Air Force. This leads to the design and manufacture of a variety of Aerostats that are still being manufactured today for use in the United States' drug interdiction efforts.

In 1970 ILC Ind. Inc. purchased Steinthal & Company Inc., an established manufacturer of military parachutes in Roxboro, N.C. The company was eventually sold in 1976.

1971 ILC Industries Inc. expands with the purchase of the Data Device Corporation located in Bohemia, NY. Data Device is engaged in the research and development, manufacture, and sale of advanced electronic components.

The Dover operation becomes known as ILC Dover, a Division of ILC Industries, Inc.

ILC Dover manufactures and markets a line of goggles used by skiers. They are known as 'ILC GOGGS".

1973 ILC Dover designs and manufactures a liquid cooling vest that is a spin-off of the cooling garment used by, the astronauts. The 'Cool Vest* m becomes an industry standard and is still being marketed by ILC Dover, Inc.

1974 Skylab program ends and ILC Dover experiences a major reduction in the work force. Over a 2-year period the work force in Delaware is down to only 25 people.

Manufacturing moves from Dover to the Frederica facility followed by engineering and technical support (tech writing, drafting, test lab, etc.).

1975 ILC Dover doses the Pear Street facility and consolidates the entire operation in Frederica.

Total operations are performed in 45,000 fl^2 of building and 2 inflatable structures. ILC rents the Frederica facility for a sum of $22,000 a year.

Under a NASA contract ILC continued space suit development activities such as Lightweight Intra-vehicular Suit, an Improved Thermal Micrometeoroid Garment, an Emergency Intravehicular Suit and an Orbital Extravehicular Suit.

ILC Dover wins a contract with the Defense Personnel support Center to manufacture goggles for military use. These goggles have multiple lenses to protect against sun, wind, and dust. ILC makes about 100,000 pairs of goggles a year for several years.

1976 ILC Dover wins a contract to develop the DPE, a highly reliable, mobile, leak-free, disposable ensemble for use in lethal chemical environments. This is for the Army's Chemical Systems Laboratory in Aberdeen, MD, now known as CBDCOM. This contract leads to our Chemturion suit line introduced in 1979.

Today, Chemturions are being used by the EPA, NIOSH, the Center for Disease Control in Atlanta, and many industrial companies such as DuPont, Dow, Georgia Pacific and Eli Lilly.

1977 ILC Dover is awarded the contract to develop, certify and manufacture the Shuttle Space Suit. ILC will subcontract to Hamilton Standard. We grow to 100 employees.

ILC Dover develops a protective garment for use by the Environmental Protective Agency Inspectors In hazardous environments such as rail, truck and industrial chemical spills.

ILC Industries, Inc. is purchased by Rapid American Corp. from Glen Alden Corp.

1978 ILC Dover further expands after being awarded a NASA contract to provide Crew Equipment and Stowage Provisions for the Space Shuttle. ILC Dover opens a new facility near Houston's Johnson Space Center, and starts a new division named ILC Space Systems.

1980 ILC Dover is awarded a contract to design and fabricate the Cyclocrane. The Cyclocrane is a hybrid aerostat with wings. Each wing has its own engine and propeller for lift and directional capability. The Cyclocrane is test flown In Oregon in 1983.

1981 ILC Dover is awarded a contract to develop collapsible fuel and water tanks for the Army. Over a 9 year period our designs of 3,000 gal. through 210,000 gal. (5,000 barrel) tanks are manufactured on an around the dock work schedule. In this 9-year period 5,280 tanks are manufactured.

We purchase the Frederica facility, which at that time consisted of Building #1 plus 35 acres of land. Funds for this purchase are obtained from the State of Delaware via an Economic Development program that provides low interest rate loans to expanding businesses within the state.

At peak employment times our company population rises to 418.

ILC Dover is awarded a contract through NASA and the Air Force to design and manufacture the Propellant Handlers Ensemble (PHE). The suits are used to protect the handlers of liquid rocket fuel. 550 suits are manufactured.

1982 Mr. Homer Reihm is named President and General Manager of ILC Dover, Inc. At this same time John McMullen and Carl Zlock are named Vice Presidents. Mr. Reihm was our Vice President and General Manager from 1976.

Rapid American sells ILC Industries to Mr. Leonard Lane. The Lane family still owns the company today.

Building 2 is erected adding 38,000 sq. ft. to our facilities.

ILC Dover is awarded the advanced development phase of the M43 Hood/Mask integrated with a battery powered motor blower filtering system, for use on the Apache helicopter.

Torpedo recovery bags are developed and manufactured through a contract with Rocket Research. Bags are used to float Navy practice torpedoes for reuse. ILC receives multiple orders for additional bags over the years we continue to receive refurbishment contracts.

1983 ILC Dover again is in the forefront of the United States space program. Our Shuttle Suit is used on STS-6 for EVA (the first flight of the Challenger) in conjunction with a satellite launch. This marks the first American to walk in space since 1974.

ILC Dover under contract from Aerojet begins its first semi-automated assembly line, which turns out 1,318,680 Air-inflated Decelerator Systems (AIDS) through 1988.

Building 3 is erected to add production floor space of 38,000 sq. ft. Our M40 mask production line is in this area now.

XM40 Development contract is won by ILC Dover to design and develop a replacement for the M17 Gas Mask.

1984 Kathy Sullivan is the first US Woman to walk in space, another milestone for ILC. She wears our Shuttle Suit.

ILC Dover becomes incorporated on June 27, 1984. Its new name is ILC Dover, Inc.

1986 Building 4 is erected adding 33,000 more sq. ft. This becomes the Balloon production area. Also Building #5 (13,500 sq. ft) is built.

ILC Dover is awarded the contract to develop an improved Collective Protective System (M28), better known internally as SCPE, for the US Army.

1987 ILC Dover wins, a contract with Boeing to finalize design and qualify the AERP Hood/Mask for Air Force use.

ILC Dover is awarded a contract from the Air Force to develop a new Transportable Collective Protective System (TCPS) for use in chemical and biological environments.

ILC begins work on the MK III advanced development space suit.

Build #6 (12,000 sq. ft) is built.

1988 ILC Dover wins production contract for the Chemical Biological Protective mask (M40). We built at a rate of 15,000 a month.

ILC is awarded a contract from General Electric for five 595K balloon systems.

1989 ILC Dover under contract from Honeywell begins production of Ram Air Decelerator (RAD) cup assemblies. Production runs for 1 year with 1,843,000 units produced.

ILC (Antigua), LTD. becomes an offshore soft-goods manufacturing facility. RADs are manufactured there along with the M40 Carriers and the HGU-41/P Hoods.

1990 ILC Dover is awarded a production contract for Collective Protection Systems (M28 and M200).

A Protective Integrated Hood Mask (PIHM) contract is awarded by Boeing to ILC Dover. 5,300 hood/masks are produced.
ILC Dover enters the world of advanced composites. We are awarded a contract from the Garret Corp. to develop spinner cones for use on jet aircraft engines.

1991 ILC Dover is awarded a contract from the US Air Force to manufacture 10,500 PIHMs.

ILC is awarded a contract by Alliant Techsystems to manufacture 100,000 heat-sealed AIDS for use on Navy Tomahawk missiles.

1992 ILC Dover is awarded a contract from Boeing to develop and fabricate the Upper Pressure Garment, Lower G Garment, and Air Cooling Garment of the F22 Aircraft Life Support System.
ILC Dover Is awarded a contract from Loral for the detailed design and fabrication of 8 each 420,000 cubic feet aerostats.

ILC Dover is awarded a contract for the partial design and full fabrication of a 620,000 cu. ft. logging balloon. Due to Its success another balloon is awarded in 1993.

ILC Dover is awarded a contract from the Air Force for the first of three 275K Aerostats.

1993 ILC Space System Division is sold to Oceaneering Space Systems, Inc. of Houston, TX-

ILC installed its first Vapor Guard Tank Cover at Relchhold Chemical Corp. in Cheswold, Delaware.

1994 ILC Dover is awarded a contract by the German company Zeppelin for the detailed design and fabrication of a hybrid airship envelope.

ILC Dover is awarded a contract from NASA's Jet Propulsion Lab (JPL) to develop and manufacture the airbag landing system for the Mars Pathfinder Mission.

ILC produces our 500,000th M40 on December 9, 1994.

1995 ILC is awarded a contract to design, develop and test an Integrated Ballistic Helmet (1131-1). This is followed by an order for 741 production units.

ILC Dover teams with FPT Industries Ltd. to be their US manufacturer of fuel cells. The fuel cells are for the V22 Osprey aircraft manufactured by Boeing Helicopter, and are required to withstand gunfire and crash impact with minimal damage.

1996 ILC wins a contract with satellite manufacturer TRW to perform preliminary development of a Space Rigidizable Antenna to enhance satellite communications systems such as cell phones and Direct TV.

ILC again begins work to put people on the moon, performing a study for the design of an Inflatable Lunar Habitat.

In November of 1996, ILC is awarded a contract to build 184,000 M40 masks.

ILC is awarded a contract to develop the Hasty Hide Shelter for the US Special forces. This camouflage system enables the individual to conceal themselves during multiple day missions.

Dec. 4th 1996 - Mars Pathfinder Mission launches from Cape Canaveral with ILC's Airbag Landing System aboard. The scheduled landing date is July 4, 1997.

www.ingramcontent.com/pod-product-compliance
Lightning Source LLC
Chambersburg PA
CBHW081820170526
45167CB00008B/3482